SURVEYING MATHEMATICS MADE SIMPLE

An original Book by

Jim Crume P.L.S., M.S., CFedS

Co-Authors
Cindy Crume
Bridget Crume
Troy Ray R.L.S.
Mark Sandwick P.L.S.

PRINTED EDITION

*****

PUBLISHED BY:

Jim Crume P.L.S., M.S., CFedS

What was that Formula?

Book 11 of this Math-Series

Copyright 2013 © by Jim Crume P.L.S., M.S., CFedS

All Rights Reserved

First publication: December, 2013

Printed by CreateSpace

Available on Kindle and other devices

Cover photo courtesy of Rick Bunger, RLS (Retired) and Shane Nauert, RLS of Geodectic Analysis, LLC
www.geodeticanalysis.com

TERMS AND CONDITIONS

The content of the pages of this book is for your general information and use only. It is subject to change without notice.

Neither we nor any third parties provide any warranty or guarantee as to the accuracy, timeliness, performance, completeness or suitability of the information and materials found or offered in this book for any particular purpose. You acknowledge that such information and materials may contain inaccuracies or errors and we expressly exclude liability for any such inaccuracies or errors to the fullest extent permitted by law.

Your use of any information or materials in this book is entirely at your own risk, for which we shall not be liable. It shall be your own responsibility to ensure that any products, services or information available in this book meet your specific requirements.

This book may not be further reproduced or circulated in any form, including email. Any reproduction or editing by any means mechanical or electronic without the explicit written permission of Jim Crume is expressly prohibited.

TABLE OF CONTENTS

INTRODUCTION..4
RIGHT TRIANGLE..5
OBLIQUE TRIANGLE..6
QUADRATIC EQUATION..8
CIRCULAR CURVES..9
SPIRAL CURVES..12
SPIRAL CURVE OFFSETS...14
VERTICAL CURVES..16
TANGENT OFFSET..18
RADIUS THROUGH THREE POINTS............................19
COMPASS RULE ADJUSTMENT...................................20
(BLM - IRREGULAR BOUNDARY ADJUSTMENT)........20
BLM - GRANT BOUNDARY ADJUSTMENT....................20
CREATE COORDINATES..21
INVERSE BETWEEN COORDINATES............................21
INTERSECTIONS...22
COORDINATE TRANSFORMATION..............................23
GEODETIC COORDINATES TO GRID COORDINATES. 26
GRID COORDINATES TO GEODETIC COORDINATES. 26
GEODETIC BEARINGS TO GRID BEARINGS..................26
GRID BEARINGS TO GEODETIC BEARINGS..................26
CONFIDENCE INTERVALS...28
COMMON SHAPES..29
CONVERSIONS..33
PHOTOGRAMMETRY..35
LEGACY FORMULAS..36
ABOUT THE AUTHOR..37

INTRODUCTION

Straight forward Step-by-Step instructions.

This book is just one part in a series of digital and printed editions on Surveying Mathematics Made Simple. The subject matter in this book will utilize the methods and formulas that are covered in the books that precede it. If you have not read the preceding books, you are encouraged to review a copy before proceeding forward with this book.

For a list of books in this series, please visit:

http://www.cc4w.net/ebooks.html

Prerequisites for this book:

A basic knowledge of geometry, algebra and trigonometry is required for the formulas shown in this book.

RIGHT TRIANGLE

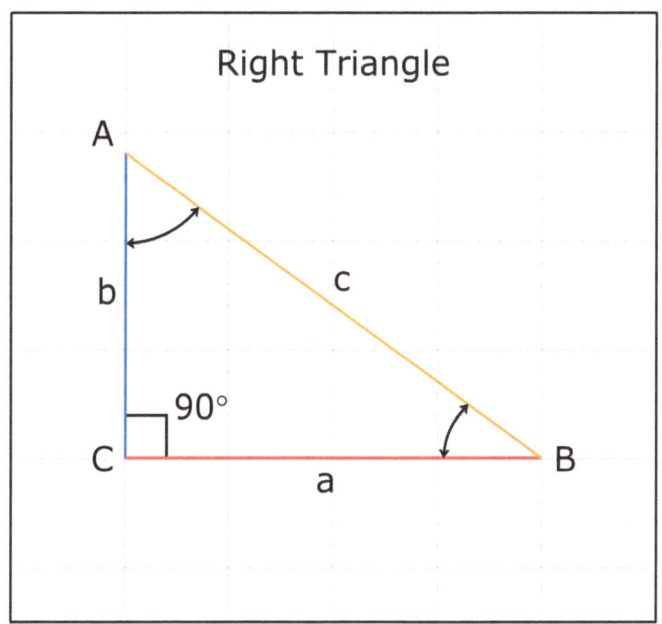

Formulas:

Area = b * a / 2

A + B + C = 180°; A + B = 90°

Internal angular closure = (N - 2) * 180

Exterior angular closure = (N + 2) * 180

$c^2 = a^2 + b^2$ ~ $c = \sqrt{(a^2 + b^2)}$ [Pythagorean Theorem]

Sin(A) = a / c ~ A = ArcSin(a / c)

Cos(A) = b / c ~ A = ArcCos(b / c)

Tan(A) = a / b ~ A = ArcTan(a / b)

Sin(B) = b / c ~ B = Arcsin(b / c)

Cos(B) = a / c ~ B = ArcCos(a / c)

Tan(B) = b / a ~ B = ArcTan(b / a)

OBLIQUE TRIANGLE

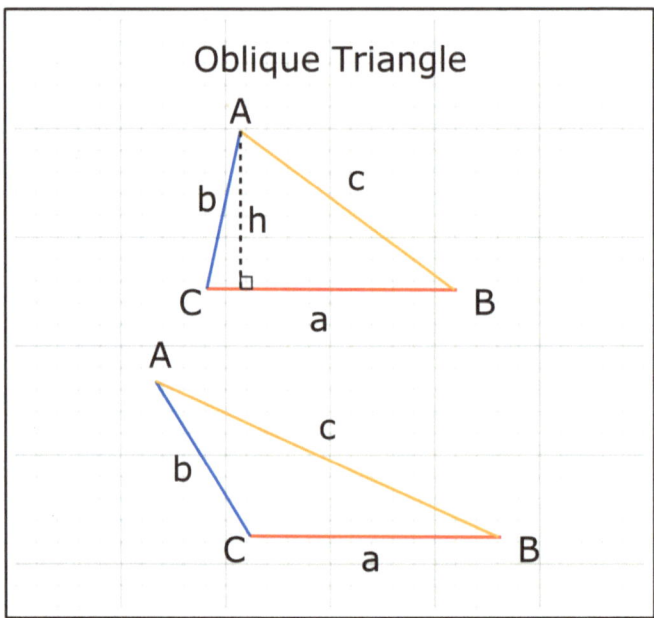

Formulas:

Area = a * h / 2

Area = Sin(C) * b * a / 2

Area = √(s * (s - a) * (s - b) * (s - c)) ~ s = (a + b + c) / 2

Area = (a² * Sin(B) * Sin(C)) / (2 * Sin(A))

A + B + C = 180°

Internal angular closure = (N - 2) * 180

Exterior angular closure = (N + 2) * 180

Law of Sins:

Sin(A) / a = Sin(B) / b = Sin(C) / c

Angles:

A = ArcSin(Sin(B) * a / b)

B = ArcSin(Sin(C) * b / c)

C = ArcSin(Sin(A) * c / a)

Sides:

$a = b * \sin(A) / \sin(B)$
$b = c * \sin(B) / \sin(C)$
$c = a * \sin(C) / \sin(A)$

Law of Cosines:

Angles:

$\cos(A) = (b^2 + c^2 - a^2) / 2 * b * c$ ~
$A = \text{ArcCos}((b^2 + c^2 - a^2) / 2 * b * c)$
$\cos(B) = (a^2 + c^2 - b^2) / 2 * a * c$ ~
$B = \text{ArcCos}((a^2 + c^2 - b^2) / 2 * a * c)$
$\cos(C) = (a^2 + b^2 - c^2) / 2 * a * b$ ~
$C = \text{ArcCos}((a^2 + b^2 - c^2) / 2 * a * b)$

Sides:

$a^2 = b^2 + c^2 - (\cos(A) * 2 * b * c)$ ~
$a = \sqrt{(b^2 + c^2 - (\cos(A) * 2 * b * c))}$
$b^2 = a^2 + c^2 - (\cos(B) * 2 * a * c)$ ~
$b = \sqrt{(a^2 + c^2 - (\cos(B) * 2 * a * c))}$
$c^2 = a^2 + b^2 - (\cos(C) * 2 * a * b)$ ~
$c = \sqrt{(a^2 + b^2 - (\cos(C) * 2 * a * b))}$

NOTES

QUADRATIC EQUATION

Formulas:

x = (-b ± √(b² - (4 * a * c))) / 2 * a

Two solutions:

x = (-b + √(b² - (4 * a * c))) / 2 * a

x = (-b - √(b² - (4 * a * c))) / 2 * a

The following form of the quadratic equation is used for solving the stationing given an elevation on a Vertical Curve:

X = (-G1 ± √((G1) ² - (2 * R * (PVCElev - Y))) / R

See **Vertical Curves** in this book and the book "**Vertical Curves - Book 10**" at http://www.cc4w.net/ebooks.html for more details on using this formula.

NOTES

CIRCULAR CURVES

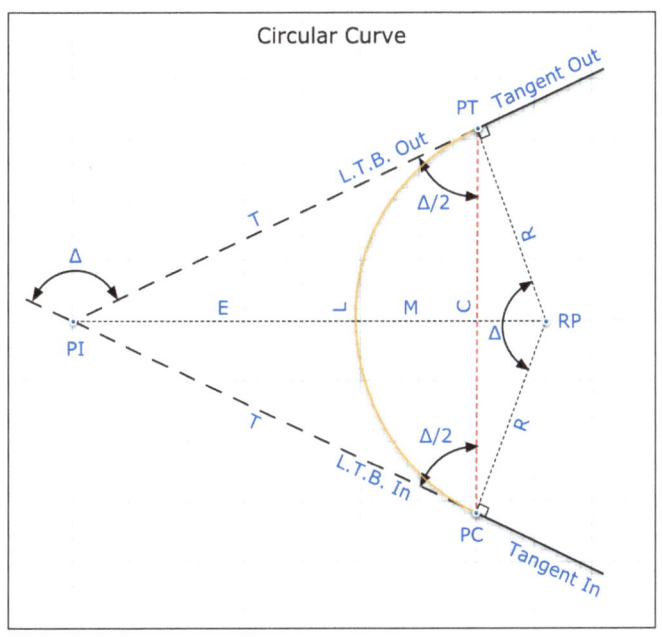

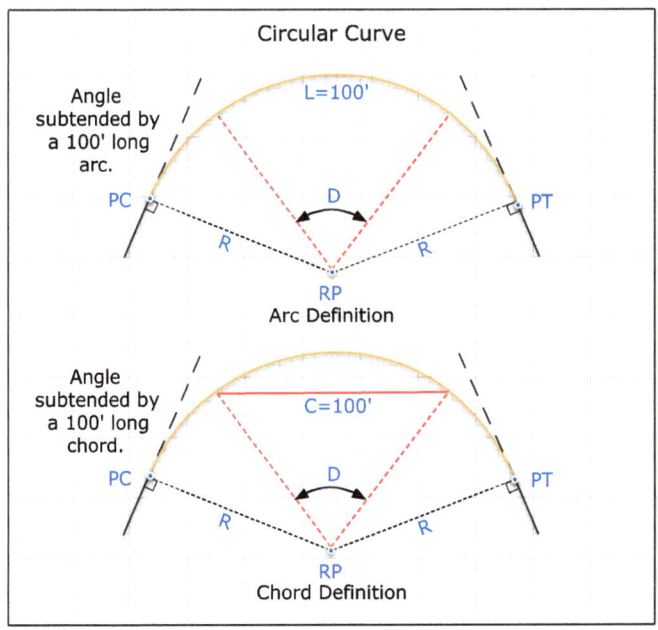

Formulas:

Arc Definition (Highway Circular Curve)

$D = 5729.57795 / R$

$R = 5729.57795 / D$

Chord Definition (Railroad Circular Curve)

$D = ArcSin(50 / R) * 2$

$R = 50 / Sin(D / 2)$

$L = 100 * \Delta / D$

$\Delta = L * D / 100$

Common to both curve definitions:

$R = (180 * L) / (\Delta * \pi)$

$R = T / Tan(\Delta / 2)$

$R = C / (2 * Sin(\Delta / 2))$

$L = (\Delta * R * \pi) / 180$

$L = (\pi * C * \Delta) / (360 * Sin(\Delta / 2))$

$L = (\pi * T * \Delta) / (180 * Tan(\Delta / 2))$

$C = 2 * R * Sin(\Delta / 2)$

$C = (360 * L * Sin(\Delta / 2)) / (\pi * \Delta)$

$C = 2 * T * Cos(\Delta / 2)$

$C = 2 * R * Sin((90 * L) / (\pi * R))$

$C = (2 * T * R) / \sqrt{(R^2 + T^2)}$

$\Delta = (180 * L) / (R * \pi)$

$\Delta = 2 * ArcTan(T / R)$

$\Delta = 2 * ArcSin(C / (2 * R))$

$T = R * Tan(\Delta / 2)$

$T = (180 * L * Tan(\Delta / 2)) / (\pi * \Delta)$

$T = C / (2 * Cos(\Delta / 2))$

$T = R * Tan((90 * L) / (\pi * R))$

$T = (R * C) / (\sqrt{((4 * R^2) - C^2)})$

$M = R * (1 - \cos(\Delta / 2))$

$M = (180 * L * (1 - \cos(\Delta / 2))) / (\pi * \Delta)$

$M = (C * \tan(\Delta / 4)) / 2$

$M = T * (1 - \cos(\Delta / 2)) / \tan(\Delta / 2)$

$E = (R / \cos(\Delta / 2)) - R$

$E = (180 * L * (1 - \cos(\Delta / 2))) / (\pi * \Delta * \cos(\Delta / 2))$

$E = (C * \tan(\Delta / 4)) / (2 * \cos(\Delta / 2))$

$E = T * \tan(\Delta / 4)$

Chord Bearing = Tangent In or L.T.B. In (+/-) ($\Delta / 2$)

Tangent Out = Tangent In or L.T.B. In (+/-) Δ

See Book 4 - **Circular Curves** - How to calculate a circular curve, reverse curve, compound curve, Tangent In, Tangent Out and Local Tangent Bearing given only two parameters. This book contains practical examples with step by step instructions and solutions for the practical examples.

NOTES

SPIRAL CURVES

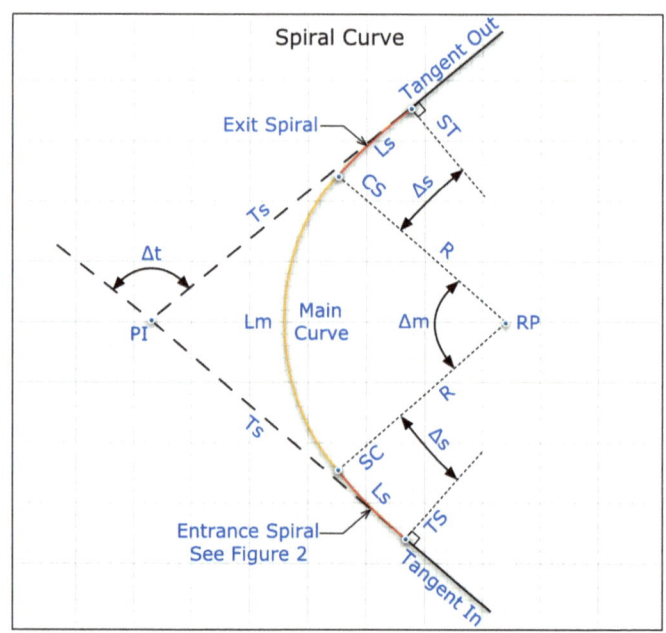

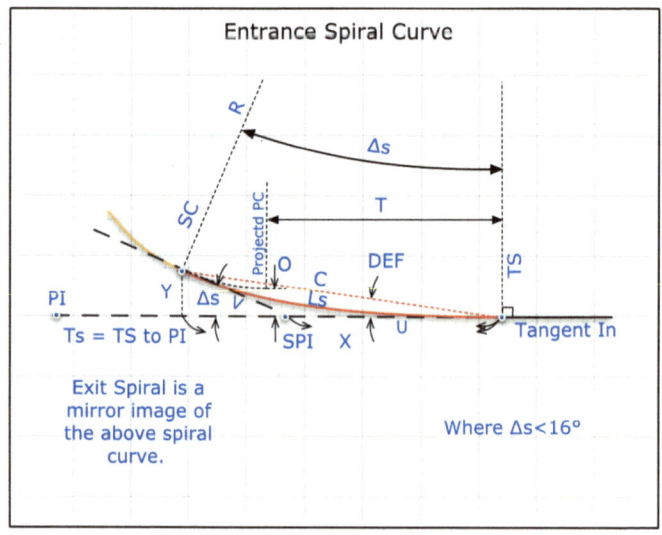

Formulas:

R = 5729.57795 / D (Highway Circular Curve)

a = (D * 100) / Ls

O = 0.0727 * a * ((Ls / 100)^3)

T = (Ls / 2) - (0.000127 * a^2 * (Ls / 100)^5)

Ts = (Tan(Δt / 2) * (R + O)) + T

C = Ls - (0.00034 * a^2 * (Ls / 100)^5)

DEF = (a * Ls^2) / 60000

Δs = 0.005 * D * Ls

U = C * Sin(Δs * 2 / 3) / Sin(Δs)

V = C * Sin(Δs * 1 / 3) / Sin(Δs)

Δm = Δt - Δs - Δs

Lm = (Δm * R * π) / 180

X = C * Cos(DEF)

Y = C * Sin(DEF)

SC(Sta) = TS(Sta) + Ls

CS(Sta) = SC(Sta) + Lm

ST(Sta) = CS(Sta) + Ls

PI(Sta) = TS(Sta) + Ts

See Book 6 - Spiral Curves - How to calculate a centerline spiral curve with minimal known information, Equal Spiral Curves, Un-Equal Spiral Curves, the mythical Ten Chord Spiral and Points on a Spiral Curve. This book contains practical examples with step by step instructions and solutions for the practical examples.

SPIRAL CURVE OFFSETS

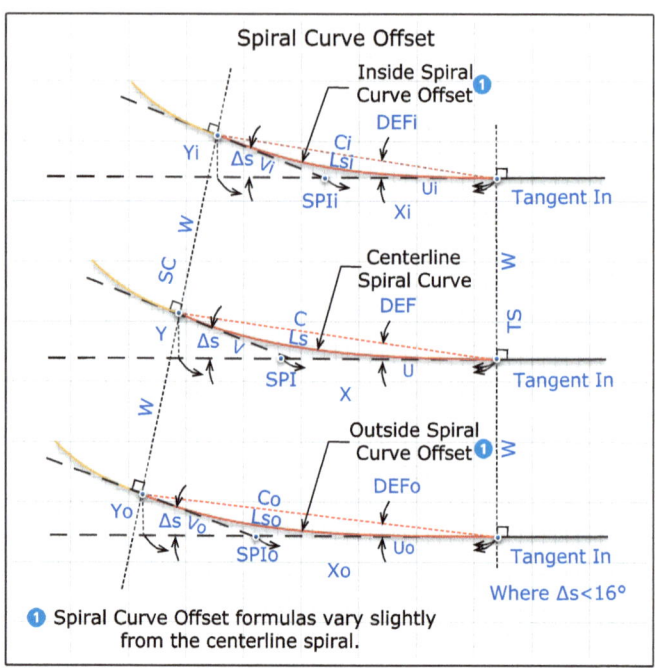

Formulas:

—Inside Spiral Curve Offset—-

$Xi = X - (Sin(\Delta s) * W)$

$Yi = Y - W + (Cos(\Delta s) * W)$

$Ci = \sqrt{(Xi^2 + Yi^2)}$

$Vi = Yi / Sin(\Delta s)$

$Ui = Xi - (Yi / Tan(\Delta s))$

$Lsi = Ci * Ls / C$

$DEFi = ArcTan(Yi / Xi)$

$Ri = R - W$

$Di = 5729.57795 / Ri$ (Arc definition)

$ai = Di * 100 / Lsi$

—Outside Spiral Curve Offset—

$Xo = X + (Sin(\Delta s) * W)$

$Yo = Y + W - (Cos(\Delta s) * W)$

$Co = \sqrt{(Xo^2 + Yo^2)}$

$Vo = Yo / Sin(\Delta s)$

$Uo = Xo - (Yo / Tan(\Delta s))$

$Lso = Co * Ls / C$

$DEFo = ArcTan(Yo / Xo)$

$Ro = R + W$

$Do = 5729.57795 / Ro$ (Arc definition)

$ao = Do * 100 / Lso$

See Book 7 - The Myth about Spiral Curve Offsets - How to calculate a true parallel spiral curve offset and points on a spiral curve offset. This book contains practical examples with step by step instructions and solutions for the practical examples.

NOTES

VERTICAL CURVES

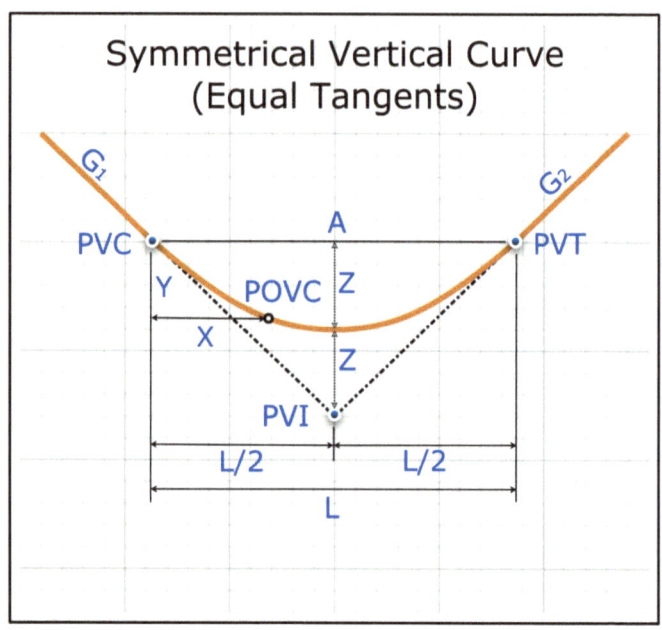

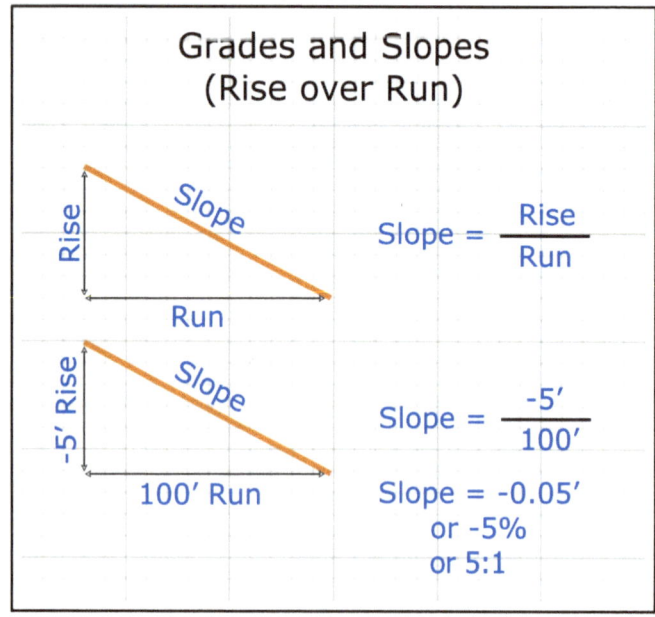

Formulas:

$A = (PVCElev + PVTElev) / 2$

$Z = (A + PVIElev) / 2$

$PVCSta = PVISta - (L / 2)$

$PVTSta = PVISta + (L / 2)$

$PVCElev = PVIElev - (G1 * (L / 2))$

$PVTElev = PVIElev + (G2 * (L / 2))$

$PVIElev$ at $POVC = (A + PVIElev) / 2$

$R = (G2 - G1) / L$ ~ *[Rate of change]*

$Y = PVCElev + G1(X) + (R / 2)X^2$ ~ *[Elevation on curve given the X value]*

$X = (-G1 \pm \sqrt{((G1)^2 - (2 * R * (PVCElev - Y)))}) / R$ ~ *[Horizontal distance along the curve given the Y value]*

$X = -G1 / R$ ~ *[High or Low point]*

See Book 10 - **Vertical Curves** - How to calculate a symmetrical vertical curve (equal tangents) and asymmetrical vertical curve (unequal tangents). This book contains practical examples with step by step instructions and solutions for the practical examples.

NOTES

TANGENT OFFSET

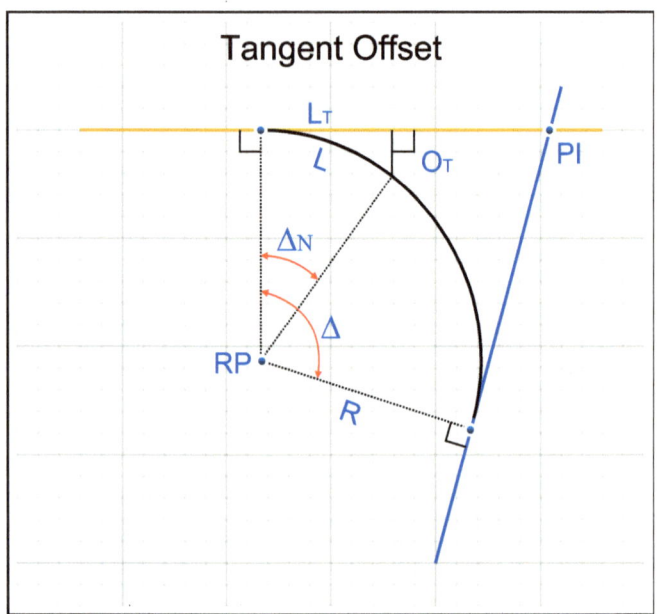

Formulas:

$\Delta N = 180 * L / \pi * R$

$O_T = R(1 - \cos(\Delta N))$

$L_T = R * \sin(\Delta N)$

O_T = Tangent Offset
L_T = Tangent Length

RADIUS THROUGH THREE POINTS

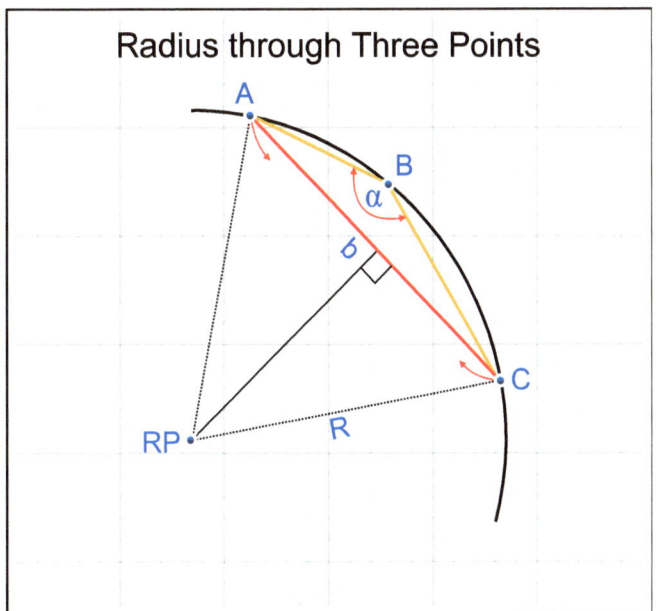

Formulas:

R = b / (2 * Sin(α))

COMPASS RULE ADJUSTMENT
(BLM - IRREGULAR BOUNDARY ADJUSTMENT)

Formulas:

P = Perimeter

ΔL = Latitude Linear Error of Closure

ΔD = Departure Linear Error of Closure

Internal angular closure = (N - 2) * 180

Exterior angular closure = (N + 2) * 180

Deflection angular closure = 360°

Error of Closure (EOC) = $\sqrt{(\Delta L^2 + \Delta D^2)}$

Closing Bearing = ArcTan(ΔD / ΔL)

Ratio of Closure = EOC / P

Adjusted Lat = Lat + (ΔL / P * (Course distance))

Adjusted Dep = Dep + (ΔD / P * (Course distance))

See Book 2 - Create Rectangular Coordinates - On how to calculate the Latitude and Departure for each course using it's bearing and distance and **Book 5** - Parcel Boundary for instructions on calculating the ΔL and ΔD Linear Error of Closure. These books contain practical examples with step by step instructions and solutions for the practical examples.

BLM - GRANT BOUNDARY ADJUSTMENT

Formulas:

Brg_1 = Record Connecting Line Bearing

Brg_2 = Measured Connecting Line Bearing

D_1 = Record Connecting Line Distance

D_2 = Measured Connecting Line Distance

Bearing Adjustment = (Course Bearing) + (Brg_2 - Brg_1)

Distance Correction = (Course Distance) * (D_2 / D_1)

CREATE COORDINATES

Formulas:

Lat (North-South direction) = Cos (Bearing*) x Distance

Dep (East-West direction) = Sin (Bearing*) x Distance

$N_2 = N_1 + Lat$

$E_2 = E_1 + Dep$

* Bearing angle must be in decimal degrees before getting the Cos or Sin value.

> **See Book 2** - **Create Rectangular Coordinates** - How to calculate the Northing and Easting of an end point given the coordinates of the beginning point, bearing and distance of a line. This book contains practical examples with step by step instructions and solutions for the practical examples.

INVERSE BETWEEN COORDINATES

Formulas:

Lat = $N_2 - N_1$

Dep = $E_2 - E_1$

Distance = $\sqrt{Lat^2 + Dep^2}$ [Pythagorean theorem]

Bearing = Arctan(Dep / Lat) [Arctangent is also known as Tan^-1]

> **See Book 3** - **Inverse between Rectangular Coordinates** - How to determine the bearing and distance of a line given the coordinates for the beginning and ending point. This book contains practical examples with step by step instructions and solutions for the practical examples.

INTERSECTIONS

Law of Cosines:

$Cos(A) = (b^2 + c^2 - a^2) / 2 * b * c$

[Solve for A as follows]

$A = ArcCos((b^2 + c^2 - a^2) / 2 * b * c)$

Once you solve for angle A then solve for the remaining two angles using the Law of Sins.

Law of Sins:

$a / Sin(A) = b / Sin(B) = c / Sin(C)$

See Book 8 - Intersections - How to calculate Bearing - Bearing, Bearing - Distance & Distance - Distance Intersections. This book contains practical examples with step by step instructions and solutions for the practical examples.

COORDINATE TRANSFORMATION

1) Northing and Easting Transformation
Formulas:

To determine the coordinate datum shift, take the differences of the Northing and Easting of a common point within both datums.

N_o & E_o = Original Coordinate Set

N_T & E_T = Transformed Coordinates

$\Delta N = N_T - N_o$ [At the common point]

$\Delta E = E_T - E_o$ [At the common point]

$N_{T(Pt.?)} = N_{o(Pt.?)} + \Delta N$ [Repeat for each point in the coordinate set]

$E_{T(Pt.?)} = E_{o(Pt.?)} + \Delta E$ [Repeat for each point in the coordinate set]

2) Elevation Transformation
Formulas:

To determine the elevation shift, take the differences of the elevation of the common point between both datum's.

EL_o = Original Coordinate Set

EL_T = Transformed Coordinates

$\Delta EL = EL_T - EL_o$ [At the common point]

$EL_{T(Pt.?)} = EL_{o(Pt.?)} + \Delta EL$ [Repeat for each point in the coordinate set]

3) Scale Transformation - Method One
Formulas:

To determine the transformed coordinate multiply the Northing and Easting of the original coordinate set by the scale factor.

N_o & E_o = Original Coordinate Set

N_T & E_T = Transformed Coordinates

$N_{T(Pt.?)} = N_{O(Pt.?)} *$ Scale Factor [Repeat for each point in the coordinate set]

$E_{T(Pt.?)} = E_{O(Pt.?)} *$ Scale Factor [Repeat for each point in the coordinate set]

4) Scale Transformation - Method Two
Formulas:

To determine the transformed coordinate first determine which original coordinate point to use as the origin.

No & Eo = Original Coordinate Set

N_T & E_T = Transformed Coordinates

$N_{T(Pt.?)} = ((N_{O(Pt.?)} - N_{ORIGIN}) *$ Scale Factor$) + N_{T\text{-}ORIGIN}$ [Repeat for each point in the coordinate set]

$E_{T(Pt.?)} = ((E_{O(Pt.?)} - E_{ORIGIN}) *$ Scale Factor$) + E_{T\text{-}ORIGIN}$ [Repeat for each point in the coordinate set]

5) Rotational Transformation - Method One
Formulas:

To determine the transformed coordinate, rotate the Northing and Easting of the original coordinate set by the angular value.

No & Eo = Original Coordinate Set

N_T & E_T = Transformed Coordinates

Distance$_{(O)(Course?)} = \sqrt{N_{O(Pt.?)}^2 + E_{O(Pt.?)}^2}$

Bearing$_{(O)(Course?)} = $ ArcTan$(E_{O(Pt.?)} / N_{O(Pt.?)})$

Bearing$_{(T)(Course?)} = $ Bearing$_{(O)(Course?)} +/-$ Angular Value

$N_{T(Pt.?)} = $ Cos(Bearing$_{(T)(Course?)}) *$ Distance$_{(O)(Course?)}$ [Repeat for each point in the coordinate set]

$E_{T(Pt.?)} = $ Sin(Bearing$_{(T)(Course?)}) *$ Distance$_{(O)(Course?)}$ [Repeat for each point in the coordinate set]

6) Rotational Transformation - Method Two
Formulas:

To determine the transformed coordinate, first determine which original coordinate point to use as the origin.

N_O & E_O = Original Coordinate Set

N_T & E_T = Transformed Coordinates

$Distance_{(O)(Course?)} = \sqrt{(N_{O(Pt.?)} - N_{ORIGIN})^2 + (E_{O(Pt.?)} - E_{ORIGIN})^2}$

$Bearing_{(O)(Course?)} = ArcTan((E_{O(Pt.?)} - E_{ORIGIN}) / (N_{O(Pt.?)} - N_{ORIGIN}))$

$Bearing_{(T)(Course?)} = Bearing_{(O)(Course?)} +/-$ Angular Value

$N_{T(Pt.?)} = N_{ORIGIN} + (Cos(Bearing_{(T)(Course?)})) * Distance_{(O)(Course?)}$
[Repeat for each point in the coordinate set]

$E_{T(Pt.?)} = E_{ORIGIN} + (Sin(Bearing_{(T)(Course?)})) * Distance_{(O)(Course?)}$
[Repeat for each point in the coordinate set]

See Book 9 - Coordinate Transformation - Coordinate Transformations that include Northing/Easting datum shift, Elevation adjustment, Scaling, Rotational, Grid to Ground, Ground to Grid and Grid Bearings to Geodetic Bearings. This book contains practical examples with step by step instructions and solutions for the practical examples.

NOTES

GEODETIC COORDINATES TO GRID COORDINATES

Formulas:

Ground Coordinate = Grid Coordinate * GAF

Note: It is very important that the GAF be applied in the correct direction. It is helpful to think of the ground distances will be longer than grid distances for most locations within the state plane zone.

GRID COORDINATES TO GEODETIC COORDINATES

Formulas:

Grid Coordinate = Ground Coordinate / GAF

Note: It is very important that the combined scale factor be applied in the correct direction. It is helpful to think of the grid distances will be shorter than ground distances for most locations within the state plane zone.

GEODETIC BEARINGS TO GRID BEARINGS

Convert Geodetic Bearing to Geodetic Azimuth

Grid Azimuth = Geodetic Azimuth - Convergence

GRID BEARINGS TO GEODETIC BEARINGS

Convert Grid Bearing to Grid Azimuth

Geodetic Azimuth = Grid Azimuth + Convergence

See Book 1 - Bearings and Azimuths - How to add bearings and angles, subtract between bearings, convert from degrees-minutes-seconds to decimal degrees, convert from decimal degrees to degrees-minutes-seconds, convert from bearings to azimuths and convert from azimuths to bearings. This book contains practical examples with step by step instructions and solutions for the practical examples.

See Book 9 - Coordinate Transformation - Coordinate Transformations that include Northing/Easting datum shift, Elevation adjustment, Scaling, Rotational, Grid to Ground, Ground to Grid and Grid Bearings to Geodetic Bearings. This book contains practical examples with step by step instructions and solutions for the practical examples.

NOTES

CONFIDENCE INTERVALS

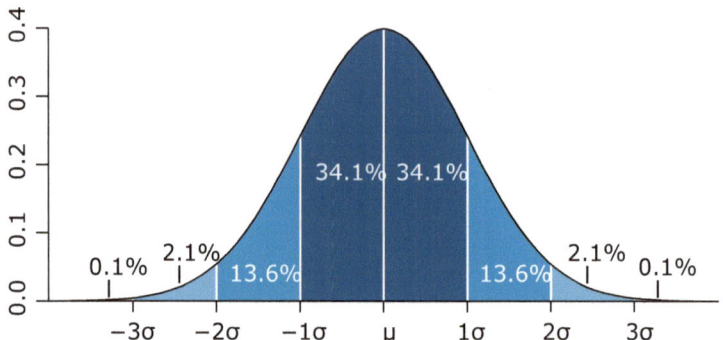

Formulas:

N = Reading #

R_N = Reading value

$M = \sum R_N / N$

$V_N = R_N - M$ ~ Repeat for each reading value

$\sigma = \sqrt{\sum V_N^2 / (N-1)}$ ~ Standard Deviation

Probability of measurement

Percentage distribution based upon Standard Deviation

68.2% ~ ±1.0 ~ -1σ to 1σ

95.4% ~ ±2.0 ~ -2σ to 2σ

99.6% ~ ±3.0 ~ -3σ to 3σ

Upper Limit

68.2% ~ M + (1.0 * σ)

95.4% ~ M + (2.0 * σ)

99.6% ~ M + (3.0 * σ)

Lower Limit

68.2% ~ M - (1.0 * σ)

95.4% ~ M - (2.0 * σ)

99.6% ~ M - (3.0 * σ)

COMMON SHAPES

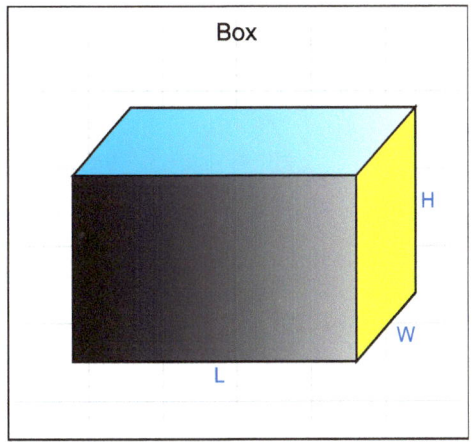

Formulas:

V = L * W * H

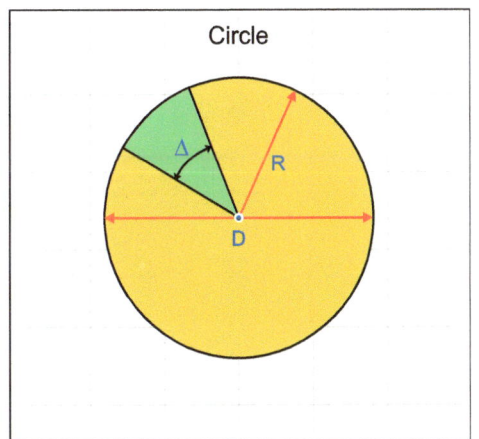

Formulas:

R = D / 2

D = R * 2

C = π * D

A = π * R²

A(Sector) = Δ / 360 * π * R²

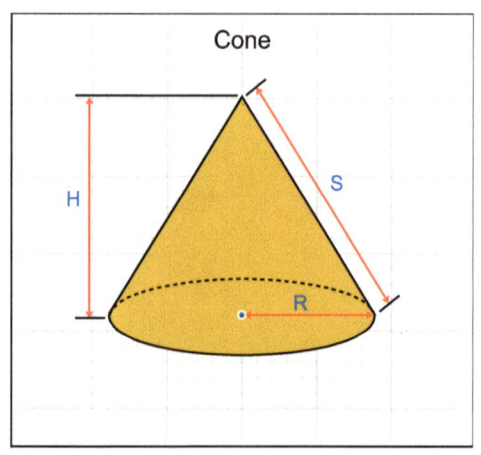

Formulas:

L = π * R * S ~ *Lateral Area*

S = (π * R^2) + (π * R * S) ~ *Surface Area*

V = 1 / 3 * π * R^2 * H

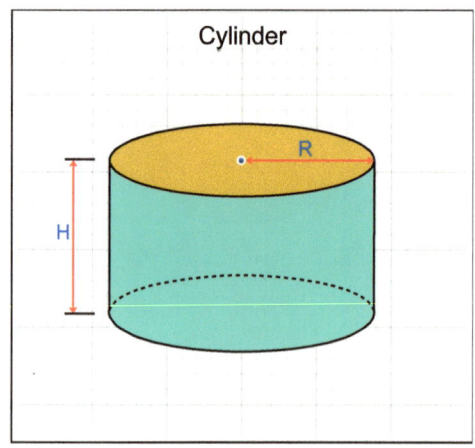

Formulas:

L = 2 * π * R * H ~ *Lateral Area*

S = (2 * π * R * H) + (2 * π * R^2) ~ *Surface Area*

V = π * R^2 * H

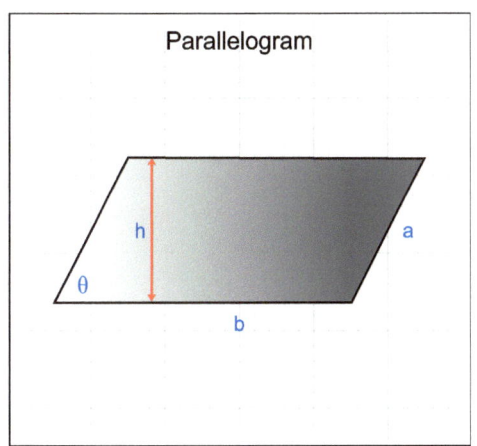

Formulas:

A = b * h

A = a * b * Sin(θ)

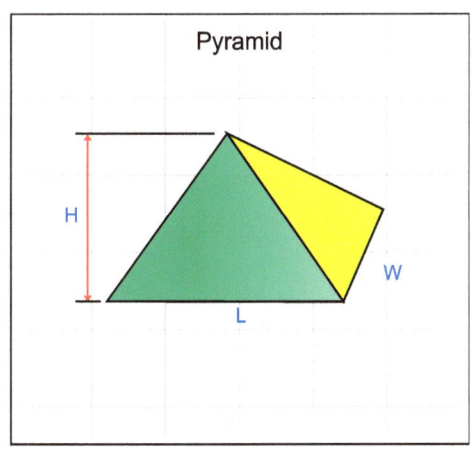

Formulas:

V = 1/3 * L * W * H

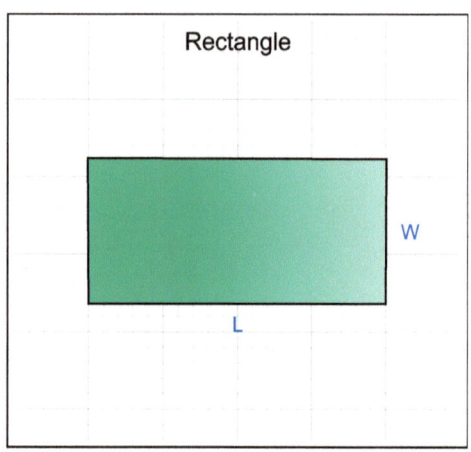

Formulas:

P = (2 * L) + (2 * W)

A = L * W

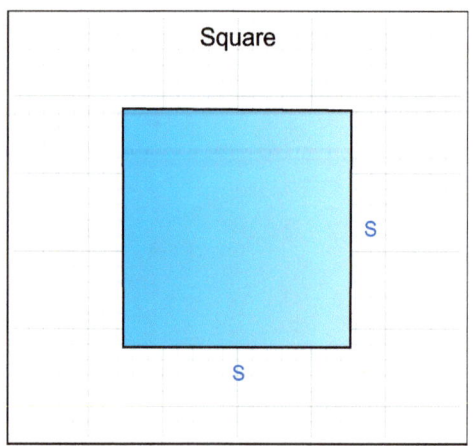

Formulas:

P = 4 * S

A = S * S

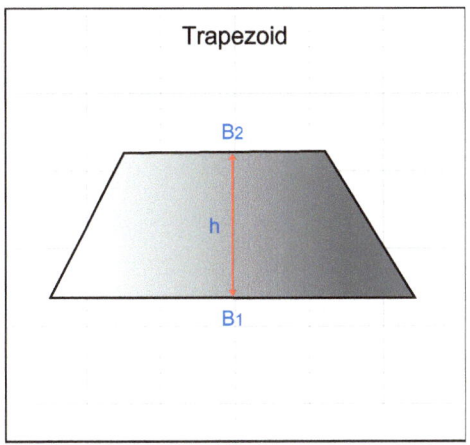

Formulas:

A = 1/2 * H * (B1 + B2)

CONVERSIONS

The conversions shown herein are those that are most commonly used in surveying.

See http://www.onlineconversion.com for a comprehensive listing of many conversions.

Distance

1 International Foot = 0.3048 meter exactly

1 Survey Foot = 0.3048006096 meter = 1200 / 3937 meter

1 inch = 1 / 12 = 0.08333 foot

1 foot = 1 / 3 = 0.33333 yard

1 yard = 3 feet

1 mile = 5280 feet

1 mile = 80 chains

1 chain = 66 feet

1 rod = 16.5 feet

1 meter = 3.280839895 feet (International)

1 meter = 3.2808333333 feet (Survey)

Area

1 sq. ft. = 144 sq. inches
1 sq. ft. = 0.11111111111 sq. yd.
1 sq. ft. (International) = 0.09290304 sq. meter
1 sq. ft. (Survey) = 0.092903412 sq. meter
1 sq. yd. = 9 sq. ft.
1 acre = 43560 sq. ft.
1 acre = 4046.8564224 sq. meter

Volume

1 acre ft. = 43560 cubic ft.
1 cubic ft. = 0.037037 cubic yd.
1 cubic ft. (International) = 0.028316846592 cubic meter
1 cubic ft. (Survey) = 0.02831701649 cubic meter
1 cubic inch = 0.0005787037037 cubic ft.
1 cubic yd. = 27 cubic ft.

Angular

1 degree = 0.0174532925199433 radian = π / 180
1 degree = 1.1111111111 grad
1 degree = 60 minutes
1 degree = 3600 seconds
1 minute = 60 seconds
1 radian = 57.2957795130823 degree = 180 / π
1 grad = 0.9 degree
45 degrees = π / 4 radians
90 degrees = π / 2 radians
135 degrees = 3 * π / 4 radians
180 degrees = π radians
225 degrees = 5 * π / 4 radians

270 degrees = 3 * π / 2 radians
315 degrees = 7 * π / 4 radians
360 degrees = 2 * π radians

PHOTOGRAMMETRY

Formulas:

f = focal length of camera in inches

H = altitude of aircraft above ground level in feet

Photo Scale = H / f

Ground Target:

Width = Photo Scale * 0.05 / 25.4

Length = 10 * Width

LEGACY FORMULAS

Legacy formulas such as Taping Corrections, Stadia Reduction, Polaris and Solar Observations have been excluded from this book. These formulas are available online and other technical writings.

Technology has advanced to a point that these legacy formulas are no longer in use. There are a few individuals that still use them in order to maintain their historic value.

I used to be the Stadia and Solar Observation guru back in the day. I have not used these methods in the past 25 years. Robotic, GPS and Scanning technology is quicker, easier and more accurate than these former technologies. They served a valuable service during their time and those who created them should be given proper recognition.

ABOUT THE AUTHOR
Jim Crume P.L.S., M.S., CFedS

My land surveying career began several decades ago while attending Albuquerque Technical Vocational Institute in New Mexico and has traversed many states such as Alaska, Arizona, Utah and Wyoming. I am a Professional Land Surveyor in Arizona, Utah and Wyoming. I am an appointed United States Mineral Surveyor and a Bureau of Land Management (BLM) Certified Federal Surveyor. I have many years of computer programming experience related to surveying.

This book is dedicated to the many individuals that have helped shape my career. Especially my wife Cindy. She has been my biggest supporter. She has been my instrument person, accountant, advisor and my best friend. Without her, I would not be the professional I am today. Cindy, thank you very much.

Other titles by this author:

http://www.cc4w.net/ebooks.html

Follow us on Facebook

Books available on Amazon.com